王大伟 著
雨青工作室 绘
[加]Jennifer May等 译

U0258584

中国水利水电出版社
www.waterpub.com.cn

小 桌 子

小 桌 子， 大 桌 子，

没 有 花 生 和 瓜 子。

扣 子 药 片 看 不 见，

光 光 溜 溜 真 好 看。

On the Table

If you have a big table or a small one?
No peanuts or seeds left out, done;
No buttons or pills that are small.
Safety first for your daughter or son.

Beware of the hidden dangers in the house.
Don't put small things like peanuts and pills on the table
because children might swallow them.

爱 爬 窗 的 小 猴 子

小 猴 子， 爱 爬 窗，

安 插 销， 窗 顶 上。

窗 边 不 放 小 板 凳，

靠 窗 不 能 放 小 床。

No Climbing

Little monkey, pledge.
Don't climb on the window ledge,
Put up bars, so I can't fall;
Off of the edge.

There should be protective bars on the windows to prevent the children from climbing or falling off from the high place.

小案例
王大伟提示

小口咬的小花猫

小花猫，小花猫，

吃东西要小口咬。

果冻可能卡嗓子，

花生瓜子不能笑。

小案例
王大伟提示

Chew your Food

I can't eat jelly, no matter my appetite.
No laughing, eating nuts, take a small bite.
Jelly can get stuck in my throat;
Always chew my food just right.

Children should try not to eat jelly. When eating snacks, children should take small bites and don't laugh because they might swallow the food and it might cause them to choke.

洗澡的绿小鸭

红小鸡，　绿小鸭，

宝宝洗澡笑哈哈。

先放凉水后热水，

不伤宝宝臭脚丫。

小案例
王大伟提示

Bath Time

It's my bath time, ready quick,
Rubber duckie and my chick,
Put cold water in first, then hot;
That's the little safety trick.

**When parents bathe their baby, put cold water
first and then hot water to avoid scalding.**

体 温 表

体 温 表， 光 闪 闪，

里 边 水 银 还 会 转。

家 有 宝 宝 藏 起 来，

含 在 嘴 里 有 危 险。

小案例
王大伟提示

Thermometer

The thermometer how it glows,
Babies can take it, and Nobody knows,
Under your arm and not in your mouth;
Out of reach and not under your nose.

**Don't let the children play with the thermometer
because if they break it the mercury
and the glass are dangerous.**

桌 上 没 有 小 药 片

胖 小 子， 胖 小 丫，

爷 爷 奶 奶 收 拾 家。

桌 上 没 有 小 药 片，

孙 子 回 来 全 家 夸 。

小案例
王大伟提示

No Pills on The Dresser

Babies, no pills to be seen.
The house is clean.
No pills on the table;
This is what I mean.

***Don't put the tablets on the table,
children might take them as snacks.***

安全小鼻孔

小汽车， 玩具熊，

掉下扣子镙丝钉，

小小零件要收好，

千万不能进鼻孔。

小案例
王大伟提示

Small Parts

No small toys for me, this is smart.
No nuts, or screws, toys fall apart,
Repair the small parts;
Take good care or they must depart.

Children might swallow or choke on toys and small parts.

桌 上 不 要 有 台 布

煤 气 灶， 热 水 瓶，

告 诉 宝 宝 不 要 动。

餐 桌 台 布 拉 下 来，

桌 上 饭 菜 盖 头 顶。

小案例
王大伟提示

No Tablecloth

No tablecloths everyone,
I could grab it and run,
Hot things on the table;
This is not fun.

Try not to put a tablecloth on the table.
Because children might grab it and pull
everything down from the table.

先 放 下 稀 饭 再 拉 孩 子

小 猴 子， 找 麻 烦，

猴 妈 妈， 有 预 案。

放 下 稀 饭 拉 孩 子，

危 机 不 乱 稳 如 山。

小案例
王大伟提示

Hot Porridge

I am little and I fall a lot.
My parents react with thought.
Put down the porridge first;
Because it is too hot.

Put down the porridge first, and then attend to the child to avoid getting hurt.

孩子面前别喝酒

孩子面前别喝酒，

宝宝误喝全家愁。

从小学习成习惯，

长大心中都是怨。

小案例
王大伟提示

No Alcohol

I won't follow; this is the key.
Adults drink in front of me.
And I know it's a bad habit;
I won't try, because I see.

It's better not to let children develop the habit of drinking.

少 糖 少 盐

少 糖 少 盐， 能 活 百 年。

不 甜 不 咸， 百 病 不 沾。

清 淡 清 淡， 最 好 习 惯。

小案例
王大伟提示

Less Sugar and Salt

I will eat light, and not ignore.
To live 100 years old or more,
Less sweet and salty food;
It's a good habit, not a chore.

Children's diet should be balanced.

两岁学会说话

小鹦鹉，　学说话，

不学坏话与脏话。

收到礼物说谢谢，

欢迎再见麻烦了。

Learning to Speak

Babies learn to talk, good childcare.
Don't say bad words, and don't swear.
Repeat thanks, your welcome, and goodbye,
These good words we can share.

***When children begin to learn to speak,
parents should beware of curse words.***

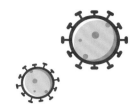

游戏篇

一、玩具小病毒

1. 玩具音量在 70 分贝以上（损伤听力）

2. 玩弓、箭、枪、飞镖等抛射性玩具

3. 玩易燃的毛绒布制玩具

4. 把玩具的包装袋套在头上，紧贴嘴巴和鼻子

5. 拆卸玩具上的小零件（易误食）

6. 玩氢气球（易爆炸）

7. 把筷子当玩具（易扎入鼻孔或戳伤眼睛）

8. 玩手机或平板电脑（易引起婴幼儿近视）

二、突发事件小病毒

1. 随便拾遗或者拆开可疑包裹（可能是炸弹）

2. 捡拾不明物体（防止接触放射源）

3. 经常带孩子去体育场馆等人多拥挤的地方

4. 让孩子独自呆在拥挤或混乱的地方，如楼梯、地下通道、桥梁等（易发生踩踏事件）

5. 在公共场所让孩子独自与陌生人玩，无家长看管

6. 在假山旁玩（石头不牢易砸伤人）

7. 在水池旁玩（易落水伤人）

8. 遇到突发事件时，孩子不及时找家长，盲目地独自解　决

三、走失拐卖小病毒
（护子银锁）

1. 居住在外来人口集中的地方

2. 家长是外来打工人口（应小心看护孩子）

3. 家长从事不稳定或繁忙的工作（应小心看护孩子）

4. 孩子小于五岁（小心看护）

5. 孩子没有固定、可靠的人看管

6. 三岁以上不会背家庭住址

7. 三岁以上不认识回家的路

8. 三岁以上不认识北斗星的方向

9. 三岁以上不会背父母的电话号码

四、突发疾病小病毒

1. 独自与小狗、小猫玩耍（易被咬伤、抓伤）

2. 被小狗、小猫抓伤只是简单包扎（应在24小时内，打狂犬疫苗针）

3. 被钉子扎破，只是简单包扎（应在24小时内，打破伤风针）

4. 发高烧时，有惊厥史的，不马上降温（防止抽风伤大脑）

5. 坐公共儿童座椅易传染孩子的独有疾病（易得小儿急疹）

图书在版编目（ＣＩＰ）数据

王大伟儿童安全童谣. 家中隐患：汉英对照 / 王大
伟著. -- 北京：中国水利水电出版社, 2021.9
ISBN 978-7-5170-9617-7

Ⅰ.①王… Ⅱ.①王… Ⅲ.①安全教育－儿童读物－
汉、英 Ⅳ.①X956-49

中国版本图书馆CIP数据核字(2021)第092324号

责任编辑	李格（1749558189@qq.com 010–68545865）	
书　　名	王大伟儿童安全童谣：家中隐患 WANG DAWEI ERTONG ANQUAN TONGYAO：JIAZHONG YINHUAN	
作　　者	王大伟　著	
绘　　图	雨青工作室	
英文翻译	[加]Jennifer May　王大伟　陈诗楠　刘原	
配音朗读	王许瞳　李晟元　郑方允　崔璎峤　侯清芸　吴郇暖　钟璇　郑淑予	
出版发行	中国水利水电出版社	
	（北京市海淀区玉渊潭南路1号D座　100038）	
	网址：www.waterpub.com.cn	
	E-mail：sales@mwr.gov.cn	
	电话：（010）68367658（营销中心）	
经　　售	北京科水图书销售中心（零售）	
	电话：（010）88383994、63202643、68545874	
	全国各地新华书店和相关出版物销售网点	
排　　版	韩雪	
印　　刷	天津久佳雅创印刷有限公司	
规　　格	210mm×190mm　24开本　5印张（总）　120千字（总）	
版　　次	2021年9月第1版　2021年9月第1次印刷	
总 定 价	68.00元（全4册）	